# Let's Talk Turkey!

by Becky Bereman Grimes

ISBN 1461078253
ISBN 978-1461078258

Published by Alternatives Unlimited
324 S. Brooks St.
Sheridan, WY 82801

Giddy-Up!
An original 2010 Leaf Art (lefart) creation by
Becky Bereman Grimes
Au@vcn.com

This book is for Susu so she knows that she is loved with all of her heart by her Aunt Becky and that is not turkey talk!

And I did all of these for her, too!

Other books by Becky Bereman Grimes

Leaf Art (lefart)
A Joyful and Playful Look at Leaves...and some poems
Leaf Art ABC...and some dances
Larry or Lariet: the adventure of a little buffalo calf
Destination: Mammoth Hot Springs for a Very Special Event
Destination: Grand Canyon of the Yellowstone
A Procession of Flowers
From Flowers to Foliage
Bird Tails
Let's Talk Turkey
and helped with A Lady in the Saddle

Did Ben really want the turkey to be the symbol of the United States of America?

In a letter to his daughter, Benjamin Franklin wrote:

For my own part I wish the Eagle had not been chosen the representative of our country. He is a bird of bad moral character. He does not get his Living honestly. You may have seen him perched on some dead tree near the river, where, too lazy to fish for himself, he watches the labor of the Fishing Hawk; and when that diligent Bird has at length taken a fish, and is bearing it to his nest for the support of his mate and young ones, the Eagle pursues him and takes it from him.

With all this injustice, he is never in good case but like those among men who live by sharping & robbing he is generally poor and often very lousy. Besides he is a rank coward: The little King Bird not bigger than a Sparrow attacks him boldly and drives him out of the district. He is therefore by no means a proper emblem for the brave and honest Cincinnati of America who have driven all the King birds from our country..."I am on this account not displeased that the figure is not known as a Eagle, but looks more like a Turkey. For the truth the Turkey is in comparison a much more respectable bird, and withal a true original native of America . . . He is besides, though a little vain & silly, a bird of courage, and would not hesitate to attack a grenadier of the British Guards who should presume to invade his farm yard with a red coat on.

This text is from Albert Henry Smyth's 1906 edition of The Writings of Benjamin Franklin, Collected and Edited with a Life and Introduction, vol. IX, page 601.

I have never lived where there were so many turkeys in town. I'm not talking about your neighbor, the gal on the phone in the car in front of you, or the guy who is blocking your driveway.

Their heavy bodies balance in the branches of the tree as the wind blows. Turkeys prefer to spend the night in tall trees, usually more than 60 feet high and situated on a ridge or at the edge of a clearing so that no obstructions will interfere with emergency flight.

They wander through the garden, yard, fields and the courthouse parking lot searching for bugs to eat. They perch on fences, shed roofs and your neighbors old swing set. There can be 20 or 30 at a time! Their backlit waddles are pretty on a January day.

The turkeys are so big that some are taller than small, compact cars which stop to let them cross. When someone beeps their horn it makes them run back and forth on or beside the road, confused about which way to go.

Wild turkeys have very powerful legs and can run at speeds up to 25 miles per hour. Even with their heavy bodies their top speed in flight is 55 miles per hour.

Heads bent the turkeys spend their day searching for food. Wild Turkeys are omnivorous, foraging for acorns as well as various seeds, berries, roots, grasses and insects. Turkeys also occasionally consume amphibians and small reptiles like snakes and salamanders. They typically forage on forest floors, but can also be found in grasslands and swamps. The gizzard is a part of a bird's stomach that contains tiny stones. It helps them grind up food for digestion. Turkeys usually feed in early morning and in the afternoon.

The male turkey, is more colorful, while the female is a beautiful brown or lighter color to camouflage her in the surroundings where she nests.

Most of the feathers of both exhibit a metallic glittering, called iridescence, with varying colors of red, green, copper, bronze and gold.

Adult male turkeys are called toms or gobblers and females are called hens. Very young birds are poults, while juvenile males are jakes and juvenile females are jennies. A group of turkeys has many collective nouns, including a "crop," "dole," "gang," "posse," "flock" and "raffle" of turkeys.

A wild turkey's gobble can be heard up to one mile away. I once watched a pair of women out walking and laughing near the turkeys. With each peal of laughter the tom turkeys would lift their heads and issue a chorus reply. Then they went back to their meal when they saw no rival.

The adult males normally weigh between 16 and 24 pounds while the females, known as hens, usually weigh between 8 and 10 pounds.

Two major characteristics distinguish males from females: spurs and beards. Both sexes have long, powerful legs covered with scales and are born with a small button spur on the back of the leg. Soon after birth, a male's spur starts growing pointed and curved and can grow to about two inches. Most hen's spurs do not grow. Gobblers also have beards, which are tufts of filaments, or modified feathers, growing out from the chest. Beards can grow to an average of 9 inches (though they can grow much longer).

The wild turkey's bald head can change color in seconds with excitement or emotion. The birds' heads can be red, pink, white or blue.

After a rainstorm turkeys shake the moisture off of their feathers like a dog shaking after a swim

Wild turkeys see in color and have excellent daytime vision that is three times better than a human's eyesight and covers 270 degrees, but they have poor vision at night.

When mating season arrives, anywhere from February to April, courtship usually begins while turkeys are still flocked together in wintering areas.

Peacocks aren't the only birds who use their fancy tails to attract a mate. Each spring male turkeys try to befriend as many females as possible. The "tom" turkeys puff up their bodies and spread their tail feathers (just like a peacock).

CARUNCLE
bumpy growths on head
and neck

SNOOD

long flap of skin,
grows from base of
beak and hangs down
over beak

WATTLE
reddish growth covering
throat and neck

USDA photo

In addition their bare head, bright snood, wattle and caruncle turn bright red when the turkey is upset or during courtship. The wattle is the flap of skin under the turkey's chin. The caruncle is brightly colored growths on the throat region.

They also perform a little turkey trot and drag their wings along the ground to make a loud sound. With that sound and their distinctive gobble they can call to female turkeys over up to a mile away.

We watched them on another evening while they were doing some courting. The tom on the left would turkey trot and drag his wing feathers around this hen. The bigger turkey in the sun ran the turkey below away from the hens.

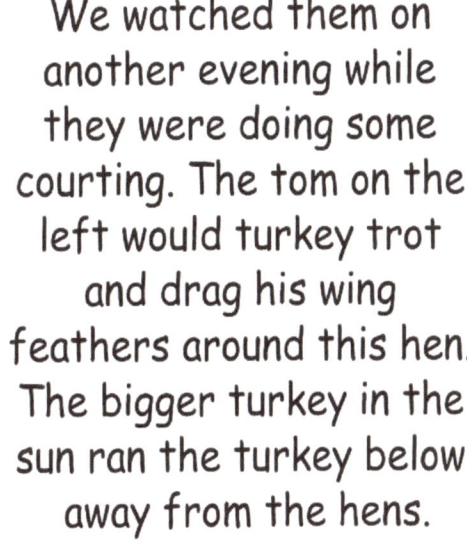

He headed around to the other side so we followed to see what he would do.

He stood there in the sunlight and unfurled his feathers and was quickly warned not to come back.

After mating, the hens begin searching for a nest site and laying eggs. In most areas, nests can be found in a shallow dirt depression, surrounded by moderately woody vegetation that conceals the nest.

Hens lay a clutch of 4 to 17 eggs in a ground nest under a bush during a two-week period. They will usually lay one egg per day. She will incubate her eggs for about 28 days, occasionally turning and rearranging them until they are ready to hatch.

Just hatched wild turkeys are precocial, which means they are born with feathers and can fend for themselves quickly, and they leave the nest within 24 hours to forage for food with their mothers. The male turkeys have very little to do with raising chicks.

At only 30,000 turkeys in the early 1900s this grand game bird was on the verge of extinction. Now there are more than 7 million turkeys. This intriguing species has truly made an awesome comeback.

We were watching them one night and didn't realize that a man and his dog were walking behind where we were sitting. This series of photographs shows how three toms protected their flock by making an even bigger threat joined together than they could do alone against the possible predator. Watch how the head, wattle and caruncle change color.

The 'what's this?' phase of alarm.

The 'run for your life!'

and "sound the alarm" phases.

The 'worried hen ' phase.

The 'this has to be taken to the counsel' phase. These three are ready to serve.

The 'something must be done!' phase.

Notice how their head color is changing? Is that because they are putting their heads together?

The "we must protect our hens!" phase.

The "new reports coming in now, sirs!" phase.

The "battle stations, men!" phase.

The "All hands on deck!" and "Keep your heads about you!" phases.

Some seem concerned
and maybe a bit confused.
Yet another decides she
is safe to eat her meal.

The heads and wattles of the counselors
are changing even more!

A single gobbler on its own is a force to be reckoned with but three is just downright scary! Look at the color changes now.

No one messes with these dudes!

With a flourish they turn and hide their flock behind their tails. The 'tom' on the left has his wing feathers down too.

As the danger passes there is a final brief show of choreographed dancing!

Then with a flourish the gobblers swirl their tails to impress the hens and, with the dog gone, the show ends.

We were entertained by the gobblers on another evening. I noticed that they liked to do their dance on the asphalt because when they drag their wings it is much louder. There was no dog to threaten them that evening and I stayed out of sight as best I could. The heads of the gobblers were a very different color that evening.

And that is the tale of the turkeys
(I'll bet you thought I was going to say the tail of a turkey.)
Put your hand on this page and draw around it. Then make
your thumb into the head and the fingers into the turkey
tail and color it to make it just the right turkey for you.

www.ingramcontent.com/pod-product-compliance
Lightning Source LLC
Chambersburg PA
CBHW050859290526
45792CB00002B/659